ABSOLUTE RELATIVITY

EQUIVALENCY OF TIME AND LIGHT SPEED AND

NATURAL RELATIVISTIC DYNAMICS OF MOVING BODIES

By

MOHSEN LUTEPHY[1]

https://www.researchgate.net/profile/Mohsen_Lutephy

---

[1] Email me at: lutephy@gmail.com

Absolute and relative are together not opposite as the black and white in Yin and Yang. Not mathematics can violate the nature and not nature violating the mathematics.

ISBN-13: 978-1091484795

b

INTRODUCTION

Einstein considered curved space and time because of his principle of light speed constancy to prove Lorentz transformation.

But here in the absolutely relativity,

The space is not curved and Galilean transformation is compatible naturally and there is no preferred coordinate system suppose the light is preferred reference.

Absolutely relativity is compatible by equivalency of the frames as the principle of relativity that,

Speed is relative but here the light speed constancy is not a principle suppose it is a proposition originated by the natural equivalency of the time and light speed allowing for mathematics to generate the formal light speed constancy which it results the Lorentz transformation absolutely and absolutely means no that the speed is measured by a preferred reference here suppose it means that the Galilean relativity is resulting the Lorentz transformation by natural equivalency of the light speed and time.

We have introduced an equivalency between Galilean transformation at the simultaneity and relativistic system at the non-simultaneity and these differences are unified here.

We have shown that the Einstein theorem of source independency of the light speed in the non-simultaneity is transferred equivalently to the source dependency of light speed in simultaneity and we argue invariance of the Maxwell equation by Lorentz transformation in the context of the source dependency of the light speed and we have found the core of the tensor covariance of the Maxwell equations and we argue the relativistic mass equation absolutely on the natural paradigms agreement naturally and also deriving straightforward argument of length contraction hypothesis by Lorentz transformation.

TABLE OF CONTENTS

LEGAL NOTES

The last chapter has been sent to a journal for publication relevant to the modification of the general relativity in a natural way.

CHAPTER 1.
ABSOLUTELY SIMULTANEITY

Einstein non simultaneity is because of law invariance and Einstein non simultaneity means that the physics is independent from existence of an absolute simultaneity.

In fact, in the Einstein relativity the physics laws are invariant in the experience and each experience is compatible by physics laws.

Then in relativity the time exists in the same face that it is experienced and the physics laws are invariant related to the same direct time experience means that there is no possible to analysis the experienced time absolutely.

Then in the Einstein relativity it is impossible to unify an instant by the physical laws in the frames and the now is not unified in the frames and in relativity, the now is not absolute suppose it is relative and it is not possible to know any relation between a now and other now in the other.

Generally, information between frames is not comparable because of the Einstein relativity that there is no any preferred reference whereas that if information between frames is comparable, it will be generated the unification between frames versus the Einstein concept of the relativity.

In the Einstein relativity each frame is an independent universe and Einstein relativity is parallel universes and parallel means that these universes are compatible but not comparable means that these universes are equivalent in the absence of the unification whereas that in the Galilean transformation,

The now is ever unified and because of the fact that in the Galilean transformation, the experience of the universe time is unified in the frames, then the now is unified in the frames and we are unified in the time in the universe by Galilean transformation and in fact twin paradox simply shows the same because that arriving after a round trip to a point results unified instantaneity that it is against the Einstein theorem.

In fact, in the Einstein relativity there is no preferred simultaneity means that simultaneity is relative and because of relative simultaneity,

What we experience it is the same existence and to the same reason the Einstein relativity agrees the principle of Bohr.

Now consider below continuum place time event that it is a sentence in the universe s that

$$x^2 - c^2 t^2 \tag{1}$$

And the same event in the universe s' that,

$$x'^2 - c^2 t'^2 \tag{2}$$

Then these sentences are equal that

$$x'^2 - c^2 t'^2 = x^2 - c^2 t^2 \tag{1-3}$$

Then against the time that it is not possible to compare it in Einstein relativity between frames, the continuum place time sentence that it is Pythagorean sentence is comparable by frames and information of continuum place time is accepted by frames as a unified information.

For a moving body the time t is compatible by relativistic equation that

$$t = \frac{t_0}{\sqrt{1 - (v/c)^2}} \tag{4}$$

We can write in the below face too that

$$t\sqrt{1 - (v/c)^2} = t_0 \tag{5}$$

$$t^2 - \left(\frac{v}{c}\right)^2 t^2 = t_0^2 \tag{6}$$

$$c^2 t^2 - v^2 t^2 = c^2 t_0^2 \tag{7}$$

For a moving body at the speed v we can use the equation x=vt in this equation and then we have

$$x^2 - c^2 t^2 = -c^2 t_0^2$$

(2-8)

But the left sentence of this equation is the same continuum place time sentence that it is compatible by equation (3) and then we obtain that

$$x'^2 - c^2 t'^2 = -c^2 t_0^2$$

(9)

But in similar way for apparatus s' we can obtain that

$$x'^2 - c^2 t'^2 = -c^2 t'^2_0$$

(10)

And then by equations (9) and (10) we obtain that

$$t'_0 = t_0$$

(11)

This is against the Einstein theorem because that it is showing unification of static clock times in different frames s and s' and then it is appeared the existence of simultaneity of times in the frames because of the fact that in the frames s and s' it is possible to unify static clocks and it is appeared unified information about the time in these frames.

On the other hand, in each frame according to the Lorentz transformation each time t is depended to the time t_0 and then when static clocks are unified then moving clocks too are unified.

In fact, if we solve the Lorentz transformation, it is appeared a function f that

$$t = f(t_0, x)$$

(12)

On the other hand, the space time continuum sentence is a Cartesian 4-space rotation and Cartesian rotation doesn't vary the meter and second for each one of the frames means that static clocks are working in these frames unified and this is equality of the meters and seconds in different frames absolutely that Einstein denied its existence for that the equality of scales in the frames results the existence of Galilean simultaneity.

Of course invariant of the scales by rotation doesn't means that the observed meter and second are invariant suppose invariant of scales here is absolutely existence but not relativistic experience and in fact each space related to a frame has the same meter and the same second that other space in other frame has it just absolutely.

This is against the Einstein theory because that according to the absolutely existence of the simultaneity, the now is absolutely unified and the persons in different frames may have the information about a unified time. in reality violation of absolute simultaneity of the frames will result the twin paradox. It is manifest that when an observer is speeding about an event then the apparatus in the event will not be changed by the speeding of the observer and then the relativistic changes are not intrinsic suppose are extrinsic and then what we see from a clock in relative speed is not what it is happened in that on point position.

In relativity you may assume that the external vison is what it exists and correlation of the frames is not a relativistic feature but we see that the correlation of the frames is not possible to deny. Even it is possible to observe experimentally that the on point observer clock is not changing by speeding of a body around that as an especial case of the twin paradox and then such an absolute unity of the intrinsic scales is observable too versus the Einstein relativity.

In fact, however relativity is experience, but experience doesn't violate intelligence.

We should notice that intelligence too does not violate experience because that observed scales from the same absolutely unified scales are different.

Absolutely equality of intrinsic scales is not observable up to now because that it needs an observer in two different frames simultaneous whereas that this is not possible actually for that in an instant an observer is just in a frame. But being at a moment in two frames simultaneous is possible absolutely because that this is intellectual and intelligent sentences are not limited to the actual impossibilities.

Of course the experience and intelligence are unified and strongly we have a tunnel to connect the absolutely simultaneity to the relatively non simultaneity.

The existence does verify experience and the experience does verify the existence whereas these are different some times.

In fact, however if it is not possible to observe station of a relatively moving frame but it is possible to consider it by intelligence.

4

In fact, the unification of scales in the frames is result of the frames equivalency and then existence of absolutely simultaneity is completely relativistic because that according to the equivalency of frames,

In each frame the clock and rule should be oneness and difference is against the equivalency of frames.

Observed moving body is possible to consider as a static body too because of the fact that an observed moving body by an on point observer is static again.

Ultimately we want to show an experimental tunnel between absolute and relative positions so that if we consider like the Einstein a lab frame and a rocket frame, according to the Lorentz transformation, observed height h is invariant in these frames and because of h constancy, there is equal constant meter h in the frames and then ever it is possible to consider a constant meter intrinsically in the frames and this is existence of absolutely unified meter in the intrinsic relativity argued experimentally.

Then in the natural relativistic dynamics of moving bodies,

Existence of equal static clocks and rules in the different frames is proved here.

The Lorentz transformation is extrinsic experience and it is possible a cause in the measurement that measurer cause to vary the intrinsic values.

For example, when light is measurer,

Light affects in the measured intrinsic scales because of the fact that measuring by light generates non simultaneity because of light traveling retardation and advances.

The same phenomenon is true for quantum physics because that uncertainty principle is not alone a mathematical evident principle suppose it is originated by the reality that when we measure the things by the light then the light momentum is an entropy itself to change the reality of the event. But we need to notice that the inability to measure the exact is not the source for nonexistence of the certainty.

Even we can consider an imaginary cause to vary the intrinsic scales extrinsically to agree with Lorentz transformation.

Mathematically in the Lorentz transformation the cause to vary intrinsic values is the Lorentz transformation because that the equation in the mathematics is the reason too.

Then in the relativity there is a universal unified clock intrinsically that the Einstein considered clocks are different experiences of the same unified clock relatively.

CHAPTER 2.
THE EQUIVALENCY OF TIME AND LIGHT SPEED

Now consider a body in the system of moving bodies from three parameters l, t, v related by equation $l = vt$.

One of these parameters is ever depended and if l is three place, the system is not five independent dimensional x, y, z, t, v suppose, the equation reduces dimension to four.

The question is that which them are depended and which one is independent?

If there is no any preference for these parameters, ever these parameters are equivalent and then if l is independent parameter, time and speed are equivalent.

Then it is resulted here that in each moving body its time and its motion are equivalent.

In the Newtonian mechanics too, there is the same phenomenon and mathematically there are three equivalencies between parameters that it is possible to generate one parameter by other and inverse, time-place and speed-time and speed-place and Kepler equations does the same three equivalencies.

Each motion and its depended time are equivalent and then it is impossible to measure a motion by its time because that this is a cycle.

Then question is that how a speed is measured?

Einstein according to the Lorentz transformation considered the clock independent of the motion to measure times in the system of moving bodies whereas that the speed depended time and its speed in a moving body are equivalent and then it is impossible to use one to other or inverse.

Like the length that it is measured by a length and measured size is a number from length per a length, the motion is too proportion and motion measured size is motion per motion.

On the other hand, impossibility of definition of each speed by its depended time causes the motion to be not measurable by its time suppose measuring should be relative that each motion should be measured by a motion in the system of moving bodies.

On the other hand, to avoid from entropy because of existence of multiple standard motions in the system of moving bodies just a motion is necessity as a reference.

In the Einstein general relativity, it is considered three variable parameters as the light speed c and time parameter t and length parameter l. But these parameters are dependent on an equation $c = dl/dt$ whereas that one of these parameters should be constant and independent and we should consider variable time and length, in a constant light speed universe or we can consider inversely the time invariant universe with variable light speed.

In relativity all things should be measurable as a Machian universe. Einstein did consider measurable values instead Newtonian absolute space and time. But what is his clock?

Length is a physical feature measurable by length and then a length is scale for length. Then we need time for time. But what is time? Is visible an independent thing in the universe we see it as the time? The length is independent but the time is completely relative measurable by both the speed and the length means $l = vt$.

Also all the speeds are possible to define by a speed and then just a speed can be considered for definition of the time and it is the light speed. This means that time is duration left by the light to travel in the length scale and it is manifest that there is no anyway.

This problem is easily describable by the mathematics for that in the algebraic it is manifest that in an equation of several parameters, one of the them is independent as a very simple case of Brouwer fixed-point theorem in mathematics and then the definition of the speed and time and meter independently on the metric tensor is not valid in general for physics.

For example, in general relativity on Schwarzschild metric we have the light variation and time variation and meter variation. But in the physics naturally light speed can be assumed constant for reality that the time is imaginary generated by the speed and it is not a true parameter like the length and then the time can be variated until the light speed to be constant and inverse.

CHAPTER 3.
THE SOURCE DEPENDENCY OF LIGHT SPEED

In a preferred frame system,

Preferred frame is reference and the existence of maximum speed is related to the preferred frame and then the maximum speed is constant related to the preferred frame.

Then the light emission at the preferred frame is at a constant speed related to the preferred frame.

In the Einstein relativity, Einstein considered source independency of light emission because of principle of light speed constancy.

In the absolutely relativity,

Relativity at the existence is returned to the simultaneity and to the same reason the emission is compatible by simultaneity.

Now question is that how the light emission is doing at the simultaneity?

We should return to the principle of relativity that,

In the relativity the speed is relative and there is no any preferred frame.

In the relativity, frames are equivalent and to the same reason,

Absolutely, the light emission is unique in each frame and difference cause to violate the equivalency of frames.

Then because of equivalency of the frames, each on point speed of light is c in each frame and this is compatible by clock invariant in each frame because that,

About the clock too the difference in on point clocks in the frames is not compatible by equivalency of the frames.

In fact, if there is not difference between frames,

Light speed emitted from each source should be invariant by each on point observer in these sources.

On the other hand, at the simultaneity the existence transformation is Galilean transformation and to the same reason intrinsically the light emission at the relativity is compatible by Galilean transformation.

To the same reason intrinsic light emission from a source observed by a relatively moving frame is compatible by linear addition of speeds because that Galilean transformation is compatible by linear addition.

Now consider two frames origins relatively moving and consider a light emitted by one of these origins and consider that emitted light is at the speed c by an on point observer.

To the same reason observed light speed c' at the intrinsic relativity by each frame is,

$$c' = c + v \qquad (13)$$

Then the light emission is not independent of the source suppose it is depended to the source at the absolutely relativity.

The intrinsic emission of the light by a source is compatible by source dependency and this results inconstancy of the light speed in the simultaneity.

Then it is resulted that,

In a relatively moving source, the light speed is not again c suppose we have

$$c' = c + v \qquad (14)$$

The intrinsic source dependency seems to opposite with Einstein independency of the light emission but the fact is that intrinsic source dependency and Einstein source independency are unified.

So to write the equations for a moving body at the simultaneity,

It is impossible to use the constancy of the light speed and if the length between points a and b is moving at the speed v related to a frame and the light is too moving in this length between the points a and b then at the simultaneity we shouldn't use below equations that

$$ab + vt_{ab} = ct_{ab}$$
$$ba - vt_{ba} = ct_{ba}$$

(15)

Because that these equations are related to the constancy of the light speed and in these equations the speed c is light speed that it is considered invariant in these equations and we should notice that,

In the ether too if we consider the light speed as a constant preferred to the ether, again the same equations are generated because of constancy of light speed in the ether.

By the way at the simultaneity,

The light speed is c+v and then at the simultaneity for the same moving light in this relatively moving length,

$$ab + vt_{ab} = (c + v)t_{ab}$$
$$ba - vt_{ba} = (c - v)t_{ba}$$

(16)

And then,

$$ab = ba = ct$$

(17)

And this is just equation for the same system absolutely.

On the other hand,

If we consider a vertical emission along the Y-axis,

Again according to the absolutely relativity that,

At the simultaneity, on point emission is moving at the speed c by each on point observer, again light speed along the Y-axis is c and in each other frame too on point observed light is moving at the speed c.

Now according to the fact that at the simultaneity naturally it is compatible the linear addition, we can generalize the light emission absolutely in the linear addition of vectors that,

11

$$\vec{c'} = \vec{c} + \vec{v} \tag{18}$$

In fact, in the Michelson morely shape of apparatus it was possible to consider an answer because of the source dependency for a considered null result and when we consider a source dependency of light emission, It is manifest that speed along the arms will be unified in a constant amount of light speed and generated result will be null again.

CHAPTER 4.
THE LIGHT SPEED CONSTANCY

4.1 THE WAY OF CONSTANCY

Einstein considered a theorem that,

Light speed constancy is a principle and it is manifest that,

Principle is improvable and to the same reason Einstein used the constancy of the light speed at the argument of the Lorentz transformation as a principle.

Then according to the Einstein relativity,

It is not possible logically to prove light speed constancy suppose especial relativity is proved on the constancy of light speed as the same process that Einstein used it to prove Lorentz transformation in his famous paper.

But is constancy of light speed a principle?

According to the equivalency of the time and light speed there is no any difference between the light speed and time of the system and according to the equivalency of the time and light speed it is possible to transfer the light speed compensated by equivalent time transformation and because of the fact that unified time t is reference then transformation of t by equivalent light speed transformation transfers the speed v depended times too because of the fact that all speed v depended times are c depended because that the c depended time t is the reference.

Then because of the equivalency of the time and light speed, equivalent time transformation does generate new equivalent system because of the fact that in its generation it is used just an equivalency means equivalency of time and light speed.

According to the equivalency of time and light speed, the time and light speed are unified together as a parameter ct that just it is possible to measure ct and according to the equivalency of the time and light speed, it is impossible to measure each of one these parameters c and t individually independent suppose just we can measure ct as a unified parameter.

Then along the time transformation, equivalently light speed should be transferred until ct to be invariant before and after time transformation.

The light speed constancy is not a principle suppose it is resulted by equivalency of the time and light speed and the time transformation does compensate transformation of the light speed until it is possible to consider a constant speed of the light at the system.

Then intrinsically the light speed c' that c'=c+v is possible to transfer to c until ct to be invariant by an equivalent time transformation that

$$(c+v)t = ct'$$ (19)

$$t' = t\frac{c+v}{c}$$ (20)

Then the light speed constancy at the generated non simultaneity because of time transformation is equivalent of absolutely light speed inconstancy at the simultaneity.

In fact, in a simultaneous system, difference at the light speeds is equivalent with light speed constancy at different times and difference at times is non simultaneity and non-simultaneity of times is time difference from a unified time.

The light speed constancy compatible by non-simultaneity is transferred to the non-constancy of light speed at the simultaneity because the equivalency of time and speed of system.

Consider lab frame and rocket frame and an apparatus of bolder and mirror in the origin of the rocket frame that b is bolder point and m is mirror point at the rocket frame and observed points are b' and m' at the lab frame and consider the light moving between bolder and mirror in this apparatus.

Light is moving ever on the points between b and m observed by on point observer and observed light by an out point observer at the lab frame is too moving on the points between b' and m'.

According to the equivalency of frames that each point in a frame is equivalent by a point in the other frame, if light is moving in the points between b and m, then observed light by lab frame is too moving on the points between b' and m'.

14

If we consider the points between b and m by an interval that $[b, m]$ then on point motion of light passing these points is a sequence ordered by < that if u and v to be two points belong to the same interval and u<v then the u is prior in time to the v.

Because of equivalency of frames we can consider observed point mathematically by an interval that $[b', m']$ and too ordered by < that u'<v' means that if u' is observed u and v' is observed v then

$$u < v \Rightarrow u' < v'$$

(21)

If the time t' is equal to t then when light is arrived to the m in the rocket frame, the observed light from lab frame is arrived too to the m'. But If transferred time t' is bigger than t, this means that when light is arrived to the m, the observed light is not arrived to the m' and then observed light should be in a point q' that,

$$b' < q' < m'$$

(22)

Then extrinsic observation is back here depended to the intrinsic observation and this means that when transferred time t' is bigger than t, observed light is back ever in comparison with the light moving in the simultaneity.

Generally, when time is transferred, extrinsic observed light is back or forward or equal related to the intrinsic simultaneity.

The back means the past and the forward means the future compared to the existence intrinsic simultaneity of frames.

In fact, if t' is bigger, observed light is at the past and if t' is smaller, observed light is at the future in comparison with the present time that it is possible to define unified at the simultaneity.

In fact, by equivalency of time and light speed we understand the mean of time transformation because that the time transformation is equivalent of light speed transformation and then when the time is transferred until light speed to be lesser, this is going to the past and if light speed is transferred to larger amounts it is generated a non-simultaneity going to the future.

Now according to the resulted facts, the observed time is not a simultaneous time and it is a time at the past or a time at the future in the relativity.

According to the Lorentz transformation, relativistic universes are transferable together, point to point and in fact transformation of coordinate system doesn't change the existence of points and existence of the points are unified in the frames and transformation of coordinate system transfers each point to a point equivalently in the existence.

Then universe is unique absolutely whereas it is possible relativistic universes and unification of universes results that intrinsically on point motion of light between bolder and mirror in the rocket frame is unified with extrinsically out point motion of light experienced by lab frame.

Finally, relativistic observed time from a moving event is not simultaneous suppose observation from a moving event is a before picture or after picture and this means that when a clock is slow means observed event is going to the past absolutely and inverse when a clock is rapid means observed universe is going to the future absolutely.

4.2 THE CONSTANCY AND MAXIMUM ARE UNIFIED

If there is a constant speed c invariant at frames,

Consider an observer frame s that observes a moving body in the speed c' that c'>c and observes a body in speed c'' that c''<c.
Now consider a frame s' that it is moving with speed v in the opposite direction related to the frame s.
By s' observed speeds of these bodies is increased and observed c' and c'' will be increased.
It is manifest that we can consider v in a size that observed c' by s' be increased until to arrive to a size rather than constant speed c.
But going from lesser speed to the rather speed is impossible else passing from a point that observed speed is equal with invariant speed c whereas that this is impossible because equality of lesser speed and c speed needs these speeds be initially equal because of invariance of c.
Then the speed of observed body that its speed is lesser than c in the s frame, it cannot arrive to the constant speed observed by s' frame and this is impossible else that the constancy and the maximum be unified.
Then if there is an invariant related to each frame, it should be maximum too.
Then because that clock is parallel with the standard motion and standard motion here should be just max speed then the clock is max velocity and it is interest that,
Equivalency of clock and max velocity seems a code of the name, Clorcks Maxwell that seems Clock is max vel.

CHAPTER 5.

EQUIVALENCY OF CLOCK AND RULE

5.1 CLOCK AND RULE EQUIVALENCY

By equivalency of the clock and max velocity, it was appeared c constancy in the relativity that in each coordinate system s the light speed is c means that

$$x = ct \tag{23}$$

This means that this is an invariant sentence and it is not varied by transformation of coordinate system and if s to be transferred to s' again

$$x' = ct' \tag{24}$$

Then the light is a preferred reference that its speed is ever invariant and to the same reason x and t are equivalent in the system.

This means that each free parameter x is related to a free parameter t by equation x=ct that c is invariant in the frames.

Then measuring x and t that they are space depended are equivalent because of invariance of c in the frames.

Then space distance is ever light traveling distance and ever there is a proportion between light traveling time and light traveling distance.

The equivalency of time and light speed after c constancy is transferred to the equivalency of time and place and equivalently, the Equivalency of clock and rule.

It is interest that,

Einstein name seems a code by the equivalency of Space and Time because of existence ST in the Einstein.

If we consider a constant distance similar to the height h perpendicular to the motion, the light traveling in the apparatus of bolder and mirror intrinsically defines universal clock that if h to be considered a meter, light traveling time in apparatus is a second and in fact apparatus of bolder and mirror is universal clock that light is its pointer.

Because of the fact that meter and second are light depended because of equivalency of clock and light then a meter is a distance that light travels it in a second and inverse a second is a time that light travels in a meter.

In fact, because of the fact that light is preferred reference then the motion of light is fundamental in the relativity.

Light by its motion in the relativity does quantize space to the scales whereas that in the physics it is considered standard meter and standard second whereas that here the light is standard meter and standard second and it is meter and it is second and it is preferred reference.

Too when we consider x=vt, the x and t, are light depended because of c dependency of v moving bodies.

In fact, not directly c depended parameters are not scale quantized parameters and just parameters that they are directly measured by light are scale quantized and to the same reason it is not possible to use x=ct in the x=vt and this is error.

But when we are on the space and space transformation it is possible to transfer ever x and t together because of equivalency of place and time in the space.

Then special relativity does not include the clock and rule suppose just clock or just rule because that rule is perfectly measured by the clock equivalently.

In the especial relativity, meter and second are unified and there is no difference between meter and second and for a distance too it should be used second like the light year instead of meter at the cosmology.

If we have a meter at the especial relativity, it is light traveled distance in a second.

On the other hand, according to the equivalency of x and t in the space, two different points have different times compatible by x=ct.

This is compatible by a circular time axis that X-axis is pillared from center to the time circles at the T-axis.

5.2 ABOUT THE EINSTEIN PAPER AND EQUIVALENCY OF CLOCK AND RULE

Einstein did not believe to the equivalency of clock and rule in relativity whereas that if Lorentz transformation is proved by Einstein just by the light speed constancy this means that the time and place transformation should be light depended because that if they are not light depended then proving just by light constancy is impossible.

Then it is manifest that in the Einstein proof too it should be used equivalency of clock and rule and in fact Einstein has used it unknowingly.

Referring to the Einstein paper shows that Einstein has used below equation that

$$x' = x - vt \qquad (25)$$

And too he has used that

$$t = \frac{x'}{c - v} \qquad (26)$$

Now if we compose these two equations together we obtain that

$$x' = x - vt = (c - v)t \qquad (27)$$

$$x = ct \qquad (28)$$

Then Einstein has used equivalency of clock and rule because of using x=ct to generate Lorentz transformation and if it is not compatible equivalency of clock and rule, the Einstein proof is certainly error.

5.3 THE STRAIGHTFORWARD ARGUMENT OF LENGTH CONTRACTION HYPOTHESIS BY LORENTZ TRANSFORMATION

In fact, two different static points in a frame are not simultaneous but Einstein and others to prove the length contraction hypothesis have used simultaneity of different static points means simultaneity of the ends of the used rule whereas that simultaneity of different points is opposite with equivalency of clock and rule.

But here it is proved that the length contraction hypothesis not by simultaneity of static points of space.

Einstein to prove the length contraction hypothesis in the famous his paper used a condition that observed time for point R along the X-axis is zero and in fact he used from the simultaneity of observed points at the ends of the observed rod and in his paper, he considered a condition t=0 and he resulted below equation for observed length x that

$$\xi = \gamma x$$
(29)

If t=0 then according to Einstein paper

$$\tau = \gamma \left(v / c^2 \right) x$$
(30)

And if we replace x with ξ we have

$$\tau = \frac{v}{c^2} \xi$$
(31)

This equation is a relation between the place and time points in the system K(ξ, τ) and this equation shows the dependency of points to the v whereas that points of the system K are independent from a relative speed related to another system and in fact in the relativity there isn't any explanation for resulted equation $\tau = \left(v / c^2 \right) \xi$ and this failure shows that the considered condition t=0 is illogical.

The famous argument of the length contraction hypothesis published in Dover publication too is similar to the Einstein argument and used condition t'₂-t'₁=0 is in fact the same Einstein condition t=0 and to the same reason it results similar sentence that

$$t_2 - t_1 = \frac{v \left(x_2 - x_2 \right)}{c^2}$$
(32)

there is other argument too, using space time interval and it is considered that,

$$X^2 - c^2 T^2 = x^2$$
(33)

and if we cube equation $X = \gamma x$ then

$$x^2 = X^2 - \frac{v^2}{c^2}X^2$$

(34)

and if we compare this equation with interval equation (33) it results that

$$T = \frac{v}{c^2}X$$

(35)

and this equation is again the same illogical equation generated in the other arguments.

other famous argument for length contraction hypothesis is related to the two ways constancy of light speed related to the below equations, a light going and coming between bolder and mirror that

$$ct'_1 = l' + vt'_1, \, ct'_2 = l' - vt'_2$$

(36)

Such arguments are along the failure of ether detection in the measurements especially Michelson morely experiment whereas that in these arguments

$$t'_1 + t'_2 = \frac{2cl'}{c^2 - v^2}$$

(37)

And it is used a condition that

$$t'_1 + t'_2 = t'$$

(38)

This is replacing addition of discrete times with a time whereas that this is an extra condition whereas we want here to prove the length contraction hypothesis without any illogical condition.

Now to prove, like the Einstein consider lab frame s'(x',t') and rocket frame s(x,t) that rocket frame is moving from a parallax point along the increasing X-axis direction.

now consider an arbitrary static point p in the s that, op is the distance between origin point o and point p and it is manifest that op is definition of x because that x is distance of a point from the origin of frame in that frame and then

$$op = x$$

(39)

In the frame s' observed p is p' as a moving point in the frame s' so that

$$o'p' = x'$$

(40)

According to the equation of the motion in the s' considered for motion started from the parallax point, if observed op by frame s' to be considered (op)' then,

$$o'p' = (op)' + vt'$$

(41)

Then because that o'p'=x' then

$$(op)' = x' - vt'$$

(42)

By Lorentz transformation it is possible to transfer parameters in the s' to s.

s' is moving related to the s in an opposite direction related to the motion of s related to the s' and if x'-vt' is used in the equation of motion we should use below transformation,

$$x' = \gamma(x + vt)$$
$$t' = \gamma\left(t + \frac{vx}{c^2}\right)$$

(43)

If we settle these equations in the (42) we obtain

$$(op)' = (\gamma(x + vt)) - v\left(\gamma\left(t + \frac{vx}{c^2}\right)\right)$$

(44)

$$(op)' = \gamma x\left(1 - \frac{v^2}{c^2}\right)$$

(45)

$$(op)' = \frac{x}{\gamma}$$

(46)

23

According to the definition of x in the frame s that x=op then

$$\left(op\right)' = \frac{op}{\gamma}$$

(47)

If op=l and (op)'=l' we have

$$l' = \frac{l}{\gamma}$$

(48)

CHAPTER 6.
GALILEAN RELATIVITY

It is resulted that the time at the relativity intrinsically is unified that

$$t' = t$$

(49)

This means that the second intrinsically is unified in the frames.

Too like the time it is possible to unify meter and then Galilean transformation is compatible transformation at the existence of relativity.

In the Galilean transformation the speed is relative, not preferred and Galilean transformation is compatible by equivalency of frames.

According to the Galilean transformation,

$$x' = x - vt$$
$$t' = t$$

(50)

6.1 HOMOGENEITY OF GALILEAN SPACE BY EQUIVALENCY OF TIME AND LIGHT SPEED

Consider lab frame s' and rocket frame s that,

A light is moving in the frame s between bolder and mirror along the Y-axis that bolder point is on the origin of frame s.

According to the source dependency of emission in the intrinsic existence of relativity, if a light be emitted along the Y-axis in this considered apparatus of bolder and mirror in the rocket frame, intrinsic on point speed is c and according to the source dependency, observed intrinsic speed by lab frame is compatible by linear addition of vectors that,

$$\vec{c'} = \vec{c} + \vec{v}$$

(51)

And then,

$$c'^2 = c^2 + v^2 \tag{52}$$

The lab frame observer is drawing an intrinsic triangle for vertically moving light in the rocket frame that

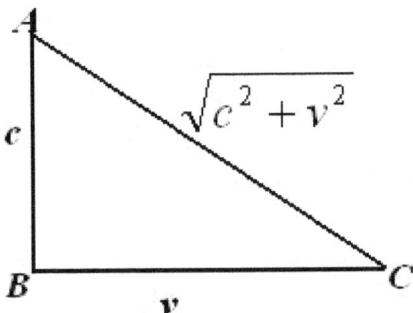

fig.1 observation of the triangle from lab frame

To generate the c constancy of light speed it needs to use from equivalency of the time and light speed to transfer time until c constancy.

Then we can use from equivalency of time and light speed to transfer c' to c and if we transfer c' to c, observed motion of light is transferred to a triangle that,

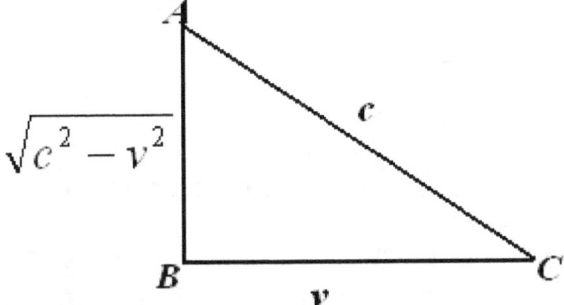

Fig.2 observation of triangle in lab frame by c constancy transformation

If c' is transferred to c, component of c' that it is c is transferred to c'ʏ that

$$c'^2_Y + v^2 = c^2 \tag{53}$$

And then,

$$c'_Y = \sqrt{c^2 - v^2}$$

(54)

The transferred time t' should be depended to the component of motion that it is along the Y-axis and then because of equivalency of time and light speed we have

$$ct = c'_Y t'$$

(55)

$$ct = \sqrt{c^2 - v^2}\, t'$$

(56)

$$t' = \frac{1}{\sqrt{1 - (v/c)^2}}\, t$$

(57)

Then two triangles are transferred together by equivalency of time and light speed that

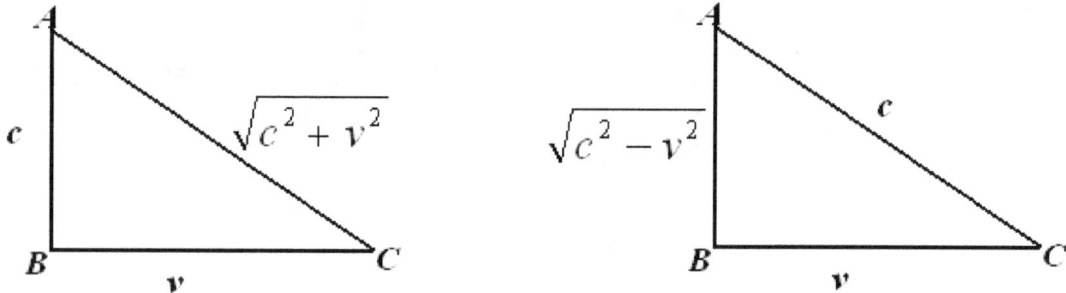

Fig.3 comparison of the triangles

These are opposite but not in conflict because that one is at simultaneity and other is at non simultaneity and these different triangles are transferable together by an equivalency resulting unification of differences.

In any way it is resulted that,

$$t' = \gamma t$$

(58)

This is time transformation for a time in the rocket frame that it is depended to a static clock that its speed is zero v=0 because that when speed is zero in the s then x=vt results x=0 and this is compatible by a static clock in the origin of rocket frame.

27

Then transferred time t' by gamma function is space depended time Independent of the motion in the rocket frame along the X-axis.

Space dependency of time transformation by gamma function is reason until because of equivalency of place and time in the space, it be possible to transfer t' and x' together by equation,

$$x' = ct' \tag{59}$$

And then

$$t' = \gamma t \rightarrow x' = \gamma x \tag{60}$$

This is homogeneity of space by gamma and if we consider K(ζ, т) that

$$\zeta = \gamma x \tag{61}$$

$$\tau = \gamma t \tag{62}$$

Then transformation between s' and K is a Galilean transformation that

$$x' = \zeta + v\tau$$
$$t' = \tau \tag{63}$$

6.2 TRANSFORMATION ALONG THE MOTION OF THE BODY

When we have considered motion of light along the Y-axis it is generated space homogeneity by gamma function.

Now consider a light moving along the X-axis in the homogeneous rocket frame K(ζ, т) that

$$\zeta = c\tau \tag{64}$$

If we settle this equation in the Galilean equation that

$$x' = \zeta + v\tau \tag{65}$$

28

It is resulted that,

$$x' = c\tau + v\tau = (c+v)\tau \tag{66}$$

And then

$$x' = (c+v)\tau$$
$$t' = \tau \tag{67}$$

According to the equivalency of time and light speed, we can transfer c+v to c constancy.

In fact, according to the equivalency of time and light speed we can equivalently transfer τ until constancy of light speed and it is manifest that it should be agreed that,

$$(c+v)\tau = ct' \tag{68}$$

We can write this equation in the below shape that

$$t' = \tau + \frac{v}{c}\tau \tag{69}$$

This equation has an extra sentence that

$$\frac{v}{c}\tau \tag{70}$$

This extra sentence causes the light speed constancy in the system of moving bodies.

On the other hand, τ is time point in the space K and then because of light speed constancy generated by the same extra sentence it is possible to transfer τ to the place point ζ by equation that,

$$\zeta = c\tau \tag{71}$$

And then it is resulted that

$$x' = \zeta + v\tau$$

$$t' = \tau + \frac{v}{c^2}\zeta$$

(73)

But K is homogenous space by gamma function from s that s(x, y) generated from equation (61,62) and then it is resulted that

$$x' = \gamma(x + vt)$$

$$t' = \gamma\left(t + \frac{v}{c^2}x\right)$$

(74)

Galilean transformation is transferred equivalently to the Lorentz transformation and using just equivalency of time and light speed shows that Galilean transformation and Lorentz transformation are equivalent so that one is intrinsic and other is extrinsic.

There is no curved space suppose experience of existence Galilean system by light constancy is difference whereas that light speed constancy is not here a principle suppose according to the equivalency of time and light speed, if time is transferred equivalently it is appeared c constancy.

Relativity is experience whereas that there is an absolute existence unified with relativity as an absolutely relativity.

The same process is compatible in the universe and we are unified together absolutely whereas we are difference relatively.

We are difference in different conditions.

A time we are bad and a time we are good and we are difference in different times and we are in the earth at simultaneity whereas we are non-simultaneous absolutely.

CHAPTER 7.
QUANTUM RELATIVITY

7.1 TRANSFORMATION ALONG THE MOTION OF THE BODY

It is generated Lorentz transformation by Galilean transformation here,

But we should notice that Lorentz transformation is an equation whereas that system of the moving bodies is included to the quantized parts too.

The twin paradox too is generated because of the fact that,

The system of the moving bodies is quantum whereas Lorentz transformation is a continuum equation.

In fact, using Lorentz transformation in the system of moving bodies splits space.

7.2 SPACE SPLIT

The coincidence of frames origins at the beginning time was an especial condition used by Einstein to generate Lorentz transformation whereas that, it is possible to occur coincidence

Not at the beginning time and to the same reason, Lorentz transformation should be generalized by an inner scalar transformation.

to the same reason in the system of moving bodies, existence of more coincidences does split the space to the quantized parts, each one separately compatible by an especial shape of Lorentz transformation.

at twin paradox too, going from a beginning coincidence after a contact is coming again to another coincidence not at the beginning and to the same reason, contact in the dynamics splits space to the separate parts and to the same reason addition of going and coming Parameters is error and this is the same error does results twin paradox.

Like the Einstein paper consider lab frame s'(x',t') and rocket frame s(x,t) that s is moving relatively related to the s' at speed v along the X-axis from a frames origins coincidence.

When frame s and frame s' origins are coincidence, coincidence means that,

frame s' observer sees origin of frame s on the origin of frame s' and to the same reason, it is evident that in the coincidence of origins,

x' in the Lorentz transformation for point x=0 should be a point that x'=0 because that the origin of frame s is a point that x=0 and origin of frame s' is a point that x'=0 and then in the coincidence of origins,

$$x = 0 \xrightarrow{\quad LT \quad} x' = 0 \tag{7-75}$$

when rocket frame s is moving along the X-axis at the speed v, it is generated the same Einstein lab frame and rocket frame and then it is appeared LT that

$$x' = \gamma\left(x + vt\right)$$

$$t' = \gamma\left(t + \frac{v}{c^2}x\right) \tag{76}$$

On the other hand,

$$\left(x = 0 \wedge x' = 0\right) \xrightarrow{\quad LT \quad} \left(t = 0 \wedge t' = 0\right) \tag{77}$$

and this means that frames s and s' are coincidence here when time is zero and this is the same Einstein preliminary consideration for lab frame and rocket frame that frames are parallax at the beginning.

now consider a wall at the static point p' on the frame s'.

Because of wall, the origin of s will arrive to the wall point p' after a time t=k and it will come back to the side of the origin s' again at speed -v and then the origin of s will arrive to the origin of s' again after a time t=k' that k'>k>0 and this is again a coincidence of origins.

we have again a coincidence of origins at time t=k' after beginning time and it should be compatible again the sentence (75) that,

$$x = 0 \xrightarrow{\quad LT \quad} x' = 0 \tag{78}$$

But according to the Lorentz transformation,

$$\left(x = 0 \wedge x' = 0 \wedge t = k'\right) \xrightarrow{\quad x' = \gamma(x+vt) \quad} \left(\gamma v k' = 0\right) \tag{79}$$

whereas that $0 = \gamma v k'$ is error.

in fact, problem is that LT is just an especial form and general form is that

$$dx' = \gamma\left(dx - vdt\right) \tag{80}$$

And then,

$$x' - x'_0 = \gamma\left((x - x_0) - v(t - t_0)\right) \tag{81}$$

and to agree with a coincidence of origins, not at the beginning time, ever there is a coefficient k' that

$$x' = \gamma\left(x - v(t - k')\right) \tag{82}$$

here we consider this kind of LT with LT* and we should notice that coincidence is not occurred at the beginning and to the same reason using LT generated by coincidence at the beginning is error whereas that using LT* shows that origins are coincidence at a time t=k' and the same fact shows that displacement of coincidence time is a time transition at the LT that it is LT* here.

here we have two faces of Lorentz transformation that one is LT for going rocket frame and one is LT* for coming rocket frame.

It is manifest that at the time of contact to the wall means t=k, these transformations are separated whereas that when time is infinitely tended to the k from lower, we should use LT and after an infinitely small time dt from contact we should use LT*.

then on the point t=k we have two faces that on the time t=k⁻,

$$x' = \lim_{t=k^-} \gamma(x - vt) \tag{83}$$

And on the time k⁺,

$$x' = \lim_{t=k^+} \gamma(x - v(t - k')) \tag{84}$$

And then if x=0,

$$\lim_{t=k^+} x' = -\gamma v(k - k') \tag{85}$$

$$\lim_{t=k^-} x' = -\gamma v k \tag{86}$$

And then it is resulted that,

$$\lim_{t=k^+} x' \neq \lim_{t=k^-} x' \tag{87}$$

and this is a discontinuity at point t=k and observed space is split on the contact time t=k and then observed space is not continuum suppose system is quantized to the two separate parts,

before contact and after contact that addition of these parts together is error.

in the relativity it is error to compose separate universes and here, before contact and after contact the systems are related to the separate relativistic universes that each one of these universes are compatible by a retarded shape of Lorentz transformation individually whereas composition is error logically.

in the twin paradox too we can consider a person moving between of a considered frame s' origin and a wall at a considered static point p' that after a time t=k it will come back to the origin of frame s' again.

in the science of relativity, it is added together observed time of going and observed coming time and then it is added values of separate universes together mistakenly and to the same reason it is appeared twin paradox in the relativity.

We should notice that one way going is the simplest shape of a system that Einstein considered it to generate below transformation that

$$x' = \gamma(x - vt)$$

$$t' = \gamma\left(t - \frac{v}{c^2}x\right)$$

(88)

but generally in the complex dynamical systems, like the accelerating systems, the condition of system is changing side by side and moment to moment and to the same reason, it should be used general shape of Lorentz transformation that,

$$dx' = \gamma(dx - vdt)$$

$$dt' = \gamma\left(dt - \frac{v}{c^2}dx\right)$$

(89)

and then without a technical problem, in each moment of system it should be considered three coefficients u,v,w and three coefficient u',v',w' that

$$x' - u = \gamma((x - v) - v(t - \omega))$$

$$t' - u' = \gamma\left((t - v') - \frac{v}{c^2}(x - \omega')\right)$$

(90)

and these coefficients are ever constant in each moment of system whereas along the time these coefficients too are variable depended to the condition of coincidence in the system.

of course we should notice that it is possible a system without a coincidence but in the same system too it is able to consider a metaphorical coincidence similar to a retarded system.

7.3 RELATIVISTIC QUANTUM

According to the source dependency of light emission at simultaneity, the light speed is compatible by linear addition of vectors that

$$\vec{c'} = \vec{c} + \vec{v}$$

(91)

In the quantum physics,

Compatibility of relativity is compatibility of source dependency of emission at simultaneity and to the same reason it is easy to realize phenomena compatible by relativity.

We should notice that,

It is possible a part of nature to be compatible by relativity and a part not.

For example, quantization of dipole momentum of fundamental particle is a relativistic feature whereas that the motion of the same particle is not.

In fact, if the speed of light was invariant related to the ether,

Dipole momentum was damping by increasing speed and for a photon dipole moment was zero whereas that,

Compatibility of relativity here causes the constancy of quantized values of dipole momentum.

7.4 NON INSTANTANEITY

According to the absolutely relativity,

Light speed is source depended and to the same reason light speed is different in the system of moving bodies absolutely.

On the other hand, it is manifest that,

Lorentz transformation is depended to the light that it is emitted from a v moving source and to the same reason mixing the lights emitted from different sources together is error in the relativity.

This means that composition of v depended Lorentz transformation by v' depended Lorentz transformation is error and in fact Lorentz transformation is in parts in the system as quantum dynamics.

In fact, composition of these frames results error and to the same reason composition of Lorentz transformations is error.

Consider three frames s" and s' and s' that s moves relatively at speed v related to s' and s' with speed u related to the s".

It is manifest that,

$$x'' = \gamma_u\left(x' + ut'\right)$$

$$t'' = \gamma_u\left(t' + \frac{u}{c^2}x'\right) \tag{92}$$

And,

$$x'' = \gamma_v\left(x + vt\right)$$

$$t'' = \gamma_v\left(t + \frac{v}{c^2}x\right) \tag{93}$$

And too,

$$x'' = \gamma_{u+v}\left(x + \frac{u+v}{1 + \dfrac{uv}{c^2}}t\right)$$

$$t'' = \gamma_{u+v}\left(t + \frac{u+v}{c^2\left(1 + \dfrac{uv}{c^2}\right)}x\right) \tag{94}$$

It is manifest that using v dependency of Lorentz transformation into the u depended Lorentz transformation is not equal to direct Lorentz transformation that it is generated because of knowledge about the addition of relativistic speed u and v.

Then it is error to compose Lorentz transformations together because of quantization of system of moving bodies.

On the other hand, if we consider a moving body in the system s that s is relatively moving related to the s' at speed v and when we consider moving body as a frame s" that x' = u t', it is appeared three frames and to the same reason generated speed by composition of Lorentz transformations for x"=0 is error and it is appeared absolutely non simultaneity.

But such non simultaneity exits because of quantization of relativity in parts.

In fact, between two frames the Lorentz transformation does generate non simultaneity that it is able to transfer to simultaneity, but between three frames generated universes are not unified and to the same reason it is impossible to unify times of three frames together.

7.5 RADIATION ABERRATION

At the simultaneity the relativistic radiation of light from a source is compatible by source dependency and to the same reason if a light is radiated from a source like a star, it will move with the motion of the mass and moving with source is character of source dependency of light emission and then in the relativity the calculation of radiation should consider that light is moving with source and then it is ever generated relativistic aberration.

For example, consider two bodies problem that light is radiated from one of these bodies and other body detects it.

CHAPTER 8.
THE 4-EQUATIONS

8.1 FUNDAMENTAL 4-EQUATIONS OF LORENTZ TRANSFORMATION

Consider two coordinate system k and k' moving relatively.

According to the Galilean transformation, if we transfer k' to k,

$$x' = x - vt \tag{95}$$

And if we transfer k to k',

$$x = x' + vt' \tag{96}$$

According to the equivalency of time and place in the space,

It is possible to use too below equations to transfer space parameters,

$$x' = ct'$$

$$x = ct \tag{97}$$

Then it is generated an apparatus of 4-equations that,

$$x = ct \tag{98}$$

$$x' = x - vt \tag{99}$$

$$x = ct \tag{100}$$

$$x = x' + vt' \tag{101}$$

This apparatus of equations is not possible to solve else we differ between x in the right side and x in the left side and for t parameter too it should be realized such difference.

Means that in the below equations,

$$x' = x - vt$$
$$x = x' + vt'$$

(102)

To solve the apparatus of four equations,

The x in an equation should not be equal to x in the other equation and in fact these parameters are related to separate transformations.

It should be realized between a parameter appeared in an equation and the same parameter appeared in the other equation.

In an equation x is observed and in the other equation x is observer and for t parameter too and to differ between these parameters we can replace right variables that,

$$x' = \bar{x} - v\bar{t}$$

(103)

$$x = \bar{x}' + v\bar{t}'$$

(104)

If we use that,

$$\bar{x} = kx$$
$$\bar{x}' = kx'$$

(105)

Then it is resulted that,

$$\bar{x} = kx \xrightarrow{x=ct,\bar{x}=c\bar{t}} \bar{t} = kt$$
$$\bar{x}' = kx' \xrightarrow{x'=ct',\bar{x}'=c\bar{t}'} \bar{t}' = kt'$$

(106)

then by these resulted sentences, two equations from equations in the (103,104) are transferred to the below equations that

$$x' = (kx) - v(kt)$$
$$x = (kx') + v(kt')$$

(107)

Then 4-equations in the (98,99,100,101) are transferred to the below 4-equations that

$$x' = ct'$$

(108)

$$x' = (kx) - v(kt)$$

(109)

$$x = ct$$

(110)

$$x = (kx') + v(kt')$$

(111)

on the other hand, if we start from the below equation that,

$$x' = k(x - vt)$$

(112)

and using equivalency of place and time to transfer space points results that,

$$x' = k\left(ct - v\frac{x}{c} \right)$$

(113)

$$t' = k\left(t - v\frac{x}{c^2} \right)$$

(114)

And then

$$x'^2 - c^2 t'^2 =$$

$$k^2 x^2 + k^2 v^2 t^2 - 2k^2 vxt - c^2 k^2 t^2 - k^2 \frac{v^2}{c^2} + 2k^2 tvx =$$

$$k^2 x^2 \left(1 - \frac{v^2}{c^2} \right) - k^2 c^2 t^2 \left(1 - \frac{v^2}{c^2} \right) - 2k^2 vxt + 2k^2 tvx =$$

$$k^2 \left(1 - \frac{v}{c^2} \right)^2 \left(x^2 - c^2 t^2 \right)$$

(115)

41

And then,

$$x'^2 - c^2t'^2 = k^2\left(1 - \frac{v}{c^2}\,^2\right)\left(x^2 - c^2t^2\right)$$

(116)

On the other hand, if instead the used equation x'=k(x-vt) we start from the below equation that

$$x = k\left(x' + vt'\right)$$

(117)

then using equivalency of place and time in the space points results that,

$$ct = k\left(ct' + v\frac{x'}{c}\right)$$

(118)

Then,

$$t = k\left(t' + v\frac{x'}{c^2}\right)$$

(119)

And then if we calculate the sentence x²-c²t² here, it is resulted like the early used calculation for sentence x'²-c²t'² that

$$x^2 - c^2t^2 = k^2\left(1 - \frac{v}{c^2}\,^2\right)\left(x'^2 - c^2t'^2\right)$$

(120)

And then if we settle the equation (116) in the equation (120) it is resulted that

$$x^2 - c^2t^2 = k^2\left(1 - \frac{v}{c^2}\,^2\right)k^2\left(1 - \frac{v^2}{c^2}\right)\left(x^2 - c^2t^2\right)$$

(121)

And then it is appeared that

42

$$1 = k^4 \left(1 - \frac{v^2}{c^2}\right)^2$$

(122)

And then,

$$k = \frac{1}{\sqrt{1 - (v/c)^2}}$$

(123)

and then using (123) in the equation (120) results that

$$x'^2 - c^2 t'^2 = x^2 - c^2 t^2$$

(124)

Then this sentence is proved here and then it is proved here that Lorentz transformation is the Cartesian rotation in the place time 4-space.

of course mathematically it is not realizable x and x or t and t or x' and x' or t' and t' in the transformation equations that a parameter be at the left or be at the right of equations and then for example, if we consider below equation that,

$$x' = k(x - vt)$$

(125)

Then using below equation that

$$x' = x'$$

(126)

And mixing equation (125) with below equations that

$$x = \gamma(x' + vt')$$

$$t = \gamma\left(t' + \frac{v}{c^2} x'\right)$$

(127)

This mixing results that,

$$x' = k\left(k(x' + vt') - vk\left(t' + \frac{v}{c^2} x'\right)\right) = k^2 x'\left(1 - \frac{v^2}{c^2}\right)$$

(128)

43

And then

$$k^2\left(1-\frac{v^2}{c^2}\right)=1$$

(129)

And then

$$k=\gamma$$

(130)

Ultimately it is proved here that Lorentz transformation is identical transformation mathematically and,

$$f\left(f^{-1}(x,y)\right)=(x,y)$$

(131)

Einstein avoided from identical transformation to prove the Lorentz transformation because of the fact that the observed parameter and observer parameter are not equal physically and to the same reason,

Einstein considered three parallel frames whereas that,

Using three frames does not change the fact that,

He has used from the identical transformation finally.

Using observer x into the observed x is an error physically but mathematics does not realize between x and x or t and t and in the calculus, not in the logic it is manifest that,

$$x=x, t=t, x'=x', t'=t'$$

(132)

8.2 WAVE INVARIANCE

We have

$$x'=x-vt$$

(133)

And the wave equation is resulted by the sentence that,

$$\varphi(x') = \varphi(x - vt)$$

(134)

Because that

$$\varphi(x') = \varphi(x - vt) \Rightarrow \nabla^2 \varphi - \frac{1}{v^2} \frac{\partial^2 \varphi}{\partial t^2} = 0$$

(135)

This wave equation is symmetric because that,

Galilean transformation is symmetric and then,

$$x = x' + vt'$$

(136)

And then two equations from four equations of 4-equations are compatible in wave that

$$x = x' + vt'$$

(137)

$$x' = x - vt$$

(138)

The wave equation is generated by Galilean transformation and then similar to the Galilean transformation it is possible to transfer time until light speed to be invariant and then

$$x = ct$$

(139)

$$x' = ct'$$

(140)

Then the wave equation is compatible by 4-equations at equations (98,99,100,101) and to the same reason wave equation is invariant by Lorentz transformation.

CHAPTER 9.
FOUR VECTORS

If we consider a central vector r that

$$r = xi + yj + zk \tag{141}$$

And we define that

$$r = vt \tag{142}$$

And if light speed is constant by frames, it is proved that the four dimensional sentence below is invariant

$$x^2 + y^2 + z^2 - c^2 t^2 \tag{143}$$

From this proposition it is resulted a generalization of 4-vectors that in each three dimensional Pythagorean space that

$$p = p_x i + p_y j + p_z k \tag{144}$$

If it is defined an equation that

$$p = vq \tag{145}$$

And maximum v to be considered constant in the system then it is generated 4-equations and below sentence

$$p^2 - c^2 q^2 \tag{146}$$

It is invariant by below transformation that

$$p' = \gamma(p - vq)$$

$$q' = \gamma\left(q - \frac{v}{c^2} p\right) \tag{147}$$

And then it is resulted that

$$p'^2 - c^2 q'^2 = p^2 - c^2 q^2$$

(148)

Then it is appeared the 4-vector that

$$\left(p_x, p_y, p_z, icq \right)$$

(149)

CHAPTER 10.
THE CORE OF COVARIANCE

Maxwell equations are generated by a covariant tensor that,

$$f_{\mu\lambda} \tag{150}$$

And Maxwell equations are generated by a derivative on the same tensor that

$$\frac{\partial f_{\mu\lambda,\lambda\mu}}{\partial\mu,\lambda} \tag{151}$$

And because invariance by derivative here, it is resulted that Maxwell equations are covariance.

But what is reason and why such equations are covariance?

10.1 THE CORE OF MAXWELL EQUATIONS COVARIANCE

The core of covariance in the Maxwell equations in the vacuum is related to the below equations that, if motion is arbitrary in an axis x then,

$$E_{y}^{'} = \gamma\left(E_{y} + v_{x}B_{z}\right) \tag{152}$$

$$B_{y}^{'} = \gamma\left(B_{y} - \frac{v_{x}}{c^{2}}E_{z}\right) \tag{153}$$

$$E_{z}^{'} = \gamma\left(E_{z} - v_{x}B_{y}\right) \tag{154}$$

$$B_{z}^{'} = \gamma\left(B_{z} + \frac{v_{x}}{c^{2}}E_{y}\right) \tag{155}$$

$$B'_y = \gamma\left(B_y + v_x E_z\right)$$

(156)

$$E'_y = \gamma\left(E_y - \frac{v_x}{c^2}B_z\right)$$

(157)

$$B'_z = \gamma\left(B_z - v_x E_y\right)$$

(158)

$$E'_z = \gamma\left(E_z + \frac{v_x}{c^2}B_y\right)$$

(159)

The field components should be zero in the x axis until Maxwell equations to be invariant in the 4-space rotation.

It is manifest that these equations are parallax with Lorentz transformation if we replace parameters with place and time parameters.

Then similar to the Lorentz transformation, we can use 4-equations to generate these electromagnetic transformations.

The core of 4-equations in the place time is related to the famous equation $x = vt$ because that two below equations are directly resulted by this equation in the simultaneity that

$$x = x' + vt'$$

(160)

$$x' = x - vt$$

(161)

And these equations are two equations of 4-equations.

In the Maxwell electromagnetism too,

Covariance of Maxwell equations should be depended to the existence of such equation that $x = vt$.

According to the electromagnetic transformations (152,153), (154,155), (156,157), (158,159) above, it is clear that for each one of these equations, there is an equation similar to the x=vt and these equations are below equations that

49

$$E_y = v_x B_z$$

$$E_z = v_x B_y$$

$$B_y = v_x E_z$$

$$B_z = v_x E_y \tag{162}$$

These equations are generated from the motion along the X-axis and if direction of motion is not preferred, these equations are generated by the vectors that,

$$E = B \times v \tag{163}$$

$$B = v \times E \tag{164}$$

Now if motion is along the X-axis, the generated equations by these vectors are that,

$$E_y = v_x B_z, E_z = v_x B_y, B_y = v_x E_z, B_z = v_x E_y \tag{165}$$

These equations are similar and to the same reason, each result for one is able to generalize for other if we replace parameters from one to other truly.

If we choice that

$$E_y = v_x B_z \tag{166}$$

Similar to the equation x=vt if we transfer the coordinate system then because of the fact that these parameters are parallax with place and time parameters in the x=vt then like the Galilean transformation,

$$E'_y = E_y + v_x B_z \tag{167}$$

And according to the symmetry in the transformation of coordinate system,

$$E_y = E'_y - v_x B'_z \tag{168}$$

and then it is appeared an apparatus that,

$$E'_y = E_y + v_x B_z \tag{169}$$

$$E_y = E'_y - v_x B'_z \tag{170}$$

Time is able to transfer until constancy of light speed in each coordinate system according to the equivalency of B_z and light speed absolutely because of equation that

$$E_y = v_x B_z \tag{171}$$

Then between two coordinate systems it is generated below equations that,

$$E_y = cB_z \tag{172}$$

$$E'_y = cB'_z \tag{173}$$

Finally it is generated below apparatus of 4-equations from these equations (169), (170), (172), (173) that

$$E_y = cB_z \tag{174}$$

$$E'_y = E_y + v_x B_z \tag{175}$$

$$E_y = E'_y - v_x B'_z \tag{176}$$

$$E'_y = cB'_z \tag{177}$$

Similar to the Lorentz transformation,

This apparatus of 4-equations is not provable else by homogeneity of electromagnetic space that

$$E'_y = \gamma \left(E_y = v_x B_z \right) \tag{178}$$

$$E_y = \gamma\left(E'_y - v_x B'_z\right) \tag{179}$$

and then like the same process started from equation (166) if we consider similar equations that

$$E_z = v_x B_y, B_y = v_x E_z, B_z = v_x E_y \tag{180}$$

It is generated below apparatus of equations that

$$E'_z = \gamma\left(E_z - v_x B_y\right) \tag{181}$$

$$B'_z = \gamma\left(B_z + \frac{v_x}{c^2} E_y\right) \tag{182}$$

$$B'_y = \gamma\left(B_y + v_x E_z\right) \tag{183}$$

$$E'_y = \gamma\left(E_y - \frac{v_x}{c^2} B_z\right) \tag{184}$$

$$B'_z = \gamma\left(B_z - v_x E_y\right) \tag{185}$$

$$E'_z = \gamma\left(E_z + \frac{v_x}{c^2} B_y\right) \tag{186}$$

And these equations are the same equations (154,155), (156,157), (158,159) and then generation of these equations shows that the core of covariance is considered by the equations (163,164).

10.2 THE MAXWELL EQUATIONS BY COVARIANCE CORE

It seems the covariance of Maxwell equations is prior to the Maxwell equations and covariance is a draft that it is transferred to the Maxwell equations finally after jump from a differential to the partial in the draft equations because that the Covariance is remained by the same jumping and to prove, consider µε=1 then

$$E = B \times v \tag{187}$$

$$B = v \times E \tag{188}$$

And if motion is arbitrary in the X-axis, we can write these equations that

$$E_x = 0, E_y = v_x B_z, E_z = -v_x B_y \tag{189}$$

$$B_x = 0, B_y = v_x E_z, B_z = -v_x E_y \tag{190}$$

If we take a differential on these equations, it is resulted that

$$dE_x = 0, dE_y = B_z dv_x + v_x dB_z, dE_z = -B_y dv_x - v_x dB_y \tag{191}$$

$$dB_x = 0, dB_y = E_z dv_x + v_x dE_z, dB_z = -E_y dv_x - v_x dE_y \tag{192}$$

Instantaneously the motion is able to consider constant and in fact the path in the motion is addition of infinite instantaneous points in a continuum system and then answer of equations in an instant of time is accepted and the value of parameters along the path is integral of these instantaneous values and then the deferential equations that they are inferred to the time points are able to write in the physical systems instantaneously generalized in the path by the calculus of integrals and then consideration of constant motion in these equations are compensated finally by the used integrals in the paths.

In any way if we consider a constant motion to generate equations instantaneously, these early equations are written in the below face,

$$dE_x = 0, dE_y = v_x dB_z, dE_z = -v_x dB_y \qquad (193)$$

$$dB_x = 0, dB_y = v_x dE_z, dB_z = -v_x dE_y \qquad (194)$$

So that,

$$v_x = \frac{dx}{dt} \qquad (195)$$

And then,

$$dE_x = 0, dE_y = \frac{dx}{dt} dB_z, dE_z = -\frac{dx}{dt} dB_y \qquad (196)$$

$$dB_x = 0, dB_y = \frac{dx}{dt} dE_z, dB_z = -\frac{dx}{dt} dE_y \qquad (197)$$

And then

$$dE_x = 0, \frac{dE_y}{dx} = \frac{dB_z}{dt}, \frac{dE_z}{dx} = -\frac{dB_y}{dt} \qquad (198)$$

$$dB_x = 0, \frac{dB_y}{dx} = \frac{dE_z}{dt}, \frac{dB_z}{dx} = -\frac{dE_y}{dt} \qquad (199)$$

Now we want to change these equations until the symmetry of equations to be invariant and keeping symmetry needs parameters positions in equations to be the same positions before and just way is the differential d to be replaced by the ∂ qualitatively.

By transfer differential d to the ∂ in the above early equations, the magnetic and electric field parameters become directly space dependent and by jump from d to the ∂, the shape of equations is invariant and if we replace differential d by ∂ in these early equations above,

$$\frac{dE_y}{dx} = \frac{dB_z}{dt}, \frac{dE_z}{dx} = -\frac{dB_y}{dt}, \frac{dB_y}{dx} = \frac{dE_z}{dt}, \frac{dB_z}{dx} = -\frac{dE_y}{dt} \qquad (200)$$

It is manifest that,

$$dE_x = 0 \Rightarrow \frac{\partial E_x}{dt} = 0 \qquad (201)$$

$$dB_x = 0 \Rightarrow \frac{\partial B_x}{dt} = 0 \qquad (202)$$

And if these two generated equations to be added to the early equations (200), these below six equations are generated that,

$$\frac{\partial E_x}{\partial t} = 0, \frac{dE_y}{dx} = \frac{dB_z}{dt}, \frac{dE_z}{dx} = -\frac{dB_y}{dt} \qquad (203)$$

$$\frac{dB_y}{dx} = \frac{dE_z}{dt}, \frac{dB_z}{dx} = -\frac{dE_y}{dt}, \frac{\partial B_x}{\partial t} = 0 \qquad (204)$$

Motion is along the x direction and then space variation is along the x and t and along the y and z is zero and from the same facts,

It is clear that if motion to be considered along the X-axis direction these six equations are the same equations if we expand below equations are generated that,

$$\frac{\partial B}{dt} = \nabla \times E \qquad (205)$$

$$\frac{\partial E}{dt} = \nabla \times B \qquad (206)$$

And these equations are Maxwell curl equations.

If instead x direction arbitrary to be selected each other direction again the Maxwell curl equations are extracted from equations of covariance core.

And then these simple symmetric equations, covariance core, are the source of Maxwell curl equations in the vacuum.

And in fact the light speed constancy is resulted by covariance core like the Maxwell equations and if we consider equations of covariance core here,

In fact the by covariance core too like the Maxwell equations it is able to generate constancy of maximum speed and if we consider equations that,

$$E = B \times v \tag{207}$$

$$B = v \times E \tag{208}$$

If we composite these equations together

$$E = v \times E \times v \tag{209}$$

$$B = v \times B \times v \tag{210}$$

And If we consider speed perpendicular to the fields,

$$E = Ev^2 \tag{211}$$

$$B = Bv^2 \tag{212}$$

And then we obtain that

$$v = 1 \tag{213}$$

And this means that there is a speed that it is constant related to any frame because vector equations are not defined related to a considered frame and then results are invariant by frames.

On the other hand, the existence of constancy and existence of the maximum is unified and to the same reason, constant speed is the maximum.

10.3 THE DIVERGENCES BY THE CURLS IN THE MAXWELL EQUATIONS

Maxwell divergences are generated by the curls and to the same reason generation of Maxwell curls by the core of covariance above is enough and to prove,

$$\nabla \times B = \frac{\partial E}{\partial t} + \rho v$$

(214)

$$\frac{\partial (\nabla \cdot E)}{\partial t} + \nabla \cdot (\rho v) = \nabla \cdot (\nabla \times B) \xrightarrow{\nabla \cdot (\nabla \times B) = 0}$$

(215)

$$\frac{\partial (\nabla \cdot E)}{\partial t} + \nabla \cdot (\rho v) = 0$$

(216)

And because of continuity equation that,

$$\frac{\partial \rho}{\partial t} = \nabla \cdot (\rho v)$$

(217)

It is resulted that,

$$\frac{\partial (\nabla \cdot E)}{\partial t} + \frac{\partial \rho}{\partial t} = 0$$

(218)

$$\nabla \cdot E = \rho + k, \frac{\partial k}{\partial t} = 0$$

(219)

The k is a time independent parameter like the density parameter and then addition of density to k introduces a repeat density parameter,

$$\rho = \rho + k$$

(220)

It was possible to consider initially that,

$$\rho = \rho + k$$

(221)

and the result of this recycle is that,

$$\nabla \cdot E = \rho \tag{222}$$

And this is extraction of a divergence by a curl in the Maxwell electromagnetism and for other divergence related to the B filed, we can use the same process if we define a zero density that,

$$\rho' = 0 \tag{223}$$

And then it is possible to write a continuity equation for the same magnetic density that,

$$\frac{\partial \rho'}{\partial t} = \nabla \cdot \left(\rho' v \right) \tag{224}$$

Then by this magnetic equation of continuity, the curl equation goes to divergence that,

$$\nabla \times B = \frac{\partial E}{\partial t} \Rightarrow \nabla \cdot B = 0 \tag{225}$$

Therefore, Maxwell four equations are reduced to the two equations because of continuity equations.

Of course some researchers think that,

The Maxwell equations are possible generated just by continuity equation and their reason is that Maxwell electrodynamics is written in the tensor that

$$\partial_\mu F^{\mu\nu} = 4\pi J^\nu \tag{226}$$

And then,

$$\partial_\nu J^\nu = 0 \tag{227}$$

And this is continuity equation and to the same reason it is supposed that inverse way too is possible that Maxwell equations are possible generated by continuity equation.

However here it is understood that curls and divergences and continuity equation are triple equivalency that it is impossible to generate one by one suppose curls are generated by continuity and diverges and divergences by curls and continuity and continuity by curls and divergences and to the same reason it is impossible to generate curls and divergences just by equation of continuity.

10.4 CHARGED EQUATIONS THE SOURCE

About the charged curl equations of Maxwell electromagnetism,

Charge is attached into the vacuum equations in the sentence ρv until the curl equations be transferred into the divergence equations by continuity equation.

If we add a sentence ρv to the vacuum equation that,

$$\nabla \times B = \frac{\partial E}{\partial t}$$

(228)

Using continuity equation does generate charged divergence equation by the charged curl equation and then charge sentence is an attachment to the vacuum equation and to the same reason the origin of vacuum equations is origin of charged equations too.

CHAPTER 11.

THE RELATIVISTIC MASS DRIVEN ABSOLUTELY

Mathematically in the relativity the proof of relativistic mass is on the equation of momentum that,

$$p = mv \tag{229}$$

In the relativity according to the equivalency of frames,

There is no any preferred frame to write momentum suppose momentum can be written relatively related to each frame equivalently and then,

If momentum is p related to a frame and the same frame is moving relatively related to another primed frame in the speed v then,

$$p' = p - vm \tag{230}$$

$$p = p' + vm' \tag{231}$$

When light speed is considered constant in the system because of equivalency of time and light speed then we can add two below equations to these early equations too as

$$p' = cm' \tag{232}$$

$$p' = cm' \tag{233}$$

And then like the place time here we have 4-equations that,

$$p = cm \tag{234}$$

$$p' = p - vm \tag{235}$$

$$p = p' + vm' \tag{236}$$

$$p' = cm' \tag{237}$$

Then like the place time these equations are compatible if,

$$p' = \gamma(p - vm)$$

(238)

$$m' = \gamma\left(m - \frac{vp}{c^2}\right)$$

(239)

in a dimension, now consider two apparatuses k and k' so that k is moving in the speed v related to the system k' and in the system k, a mass is moving in the speed −v.

Then k is moving in the speed v along the considered axis and mass is moving in the speed −v in the k.

It is clear that,

Observed speed for mass from k' will be zero and then in the k' the same mass is static and then it is static mass in the k'.

Then m' is static mass observed from k' whereas m is moving mass in the k in the speed -v and then according to the above equations,

$$p' = \gamma(p + vm)$$

(240)

$$m' = \gamma\left(m + \frac{vp}{c^2}\right)$$

(241)

And m is a mass moving in the speed -v in the k system so that,

$$p = -mv$$

(242)

And then it is resulted that

$$p' = \gamma(-mv + vm)$$

(243)

$$m' = \gamma\left(m + \frac{-vmv}{c^2}\right) \tag{244}$$

Then,

$$p' = 0 \tag{245}$$

$$m' = \gamma m\left(1 - \frac{v^2}{c^2}\right) \tag{246}$$

And then,

$$m' = \gamma m \frac{1}{\gamma^2} \tag{247}$$

$$m = \gamma m' \tag{248}$$

And it was told above that the m' is static mass and then,

$$m = \gamma m_0 \tag{249}$$

In fact, we should notice that the static mass is mass of rigid bodies and then we can prove relativistic static mass from the rigid equations directly and to prove consider a rigid 4-momentum that it is invariant rigidly that

$$p'^2 - c^2 m'^2 = p^2 - c^2 m^2 \tag{250}$$

And when speed is zero then,

$$p' = 0 \tag{251}$$

And then

$$-c^2m'^2 = p^2 - c^2m^2$$

(252)

And when speed is zero then

$$m' = m_0$$

(253)

And then

$$-c^2m_0^2 = p^2 - c^2m^2$$

(254)

And then

$$m^2v^2 - c^2m^2 = -c^2m_0^2$$

(255)

$$m^2 = m_0^2\left(\frac{1}{1-\dfrac{v^2}{c^2}}\right)$$

(256)

$$m = m_0\gamma$$

(257)

The static mass is two ways and then the mass transformation should be related to a two ways transformation that it is transformation by gamma function and to the same reason it is mistake using below transformation not at a rigid body.

$$p' = \gamma(p + vm)$$

(258)

$$m' = \gamma\left(m + \frac{vp}{c^2}\right)$$

(259)

In fact, when a light is moving vertically on a frame this vertically motion of light is equivalent with a two ways moving of light in the same frame and then if we consider a light vertically moving in the rocket frame then the speed of light absolutely is

63

compatible with linear addition and then for a static mass that it is as a light vertically moving, its observed speed by lab frame is that

$$c'^2 = \sqrt{c^2 + v^2}$$
(260)

Then mass transformation by gamma function here is indeed related to the constancy of two ways light speed and mass transformation in the relativity should be along the same process.

But what is relation between mass and speed until mass is changed by transformation of two ways light speed?

The proof is based on the momentum equation p=mv because that according to the below equation that

$$f = \frac{d(m\vec{v})}{dt}$$
(261)

If force is parallel with direction of motion, it is manifest that force is not able to vary the vertical amount of a momentum and just it is able to vary the horizontal momentum and then if we consider a vertically moving light in the rocket frame,

Then its momentum should be invariant by the parallel force and then when a parallel force causes to move the rocket frame at the speed v related to the lab frame than,

$$m'c' = mc$$
(262)

It is manifest that on point speed c of vertically moving light is being observed extrinsically by a v moving frame that,

$$c'^2 = \sqrt{c^2 + v^2}$$
(263)

But after c constancy it is manifest that the extrinsically observed speed of vertically moving light is c'$_Y$ that,

$$c'^2_Y = \sqrt{c^2 - v^2}$$
(264)

And then because of momentum conservation along the Y-axis by parallel force then,

$$m\sqrt{c^2 - v^2} = m_0 c \tag{265}$$

And then,

$$m = m_0 \gamma \tag{266}$$

We can see that the definition of clock and definition of mass are unified and clock and mass are box that light is moving in the inside of this box and simply we can consider a bolder and mirror apparatus to define clock and too the mass.

But question is that how mass is really transferred?

In fact, light speed constancy is not a principle absolutely and mass transformation needs here a natural reason.

But what is that?

In fact, when mass is transferred by gamma function it is reason to transfer light speed by gamma function and because of equivalency of time and light speed it is reason to transfer time by gamma function that,

$$t = \frac{t_0}{\sqrt{1 - \dfrac{v^2}{c^2}}} \tag{267}$$

And when this transformation is transformation of vacuum space then because of equivalency of place and time in the vacuum space,

$$x = \frac{x_0}{\sqrt{1 - \dfrac{v^2}{c^2}}} \tag{268}$$

And then mass transformation reasons homogeneity of space by gamma function naturally.

Then it remains transformation of mass by gamma function because that light speed constancy is not principle here suppose mass transformation cause to c constancy when mass is transferred by gamma function.

REFERENCES

[1] A. A. Michelson and E. W. Morley (1887),

A. A. Michelson and E. W. Morley, Am. J. Sci. 34 333 (1887)

(john Wiely and sons. ISBN 0-471-87373-X)

[2] Einstein, albert (1905),

"On the Electrodynamics of Moving Bodies", Annalen der Physik 322 (10): 891–921, Bibcode 1905AnP...322..891E,doi:10.1002/andp.19053221004 .

[3] von Laue, Max (1911/2). "Two Objections Against the Theory of Relativity and their Refutation". *Physikalische Zeitschrift* 13: 118–120.

[4] Born, Max (1964),

Einstein's Theory of Relativity

(dover publications, ISBN 0486607690, 1964).

[5] Einstein (1911),

Annalen der Physik 322 (10): 891–921, Bibcode 1905AnP...322..891E,doi:10.1002/andp.19053221004 . }

About The Author
MOHSEN LUTEPHY

I am a scientist as a researcher on the different fields of the science especially on the fields of the mathematics and physics and earth and life sciences and I am developing the true way of the science and scientific society and I am not working on the null hypotheses suppose my books and papers are on the proofs and the facts and I have many books unpublished yet and I will try to publish them too.

visit (https://www.researchgate.net/profile/Mohsen_Lutephy)

OTHER BOOKS BY (AUTHOR)

List your other kindle books with a link to the page

https://www.amazon.es/Machian-Mond-Modification-Dynamics-Principle/dp/1519144024

https://www.amazon.co.uk/absolute-physics-non-scale-mechanics/dp/1492164216

https://www.amazon.com/absolute-dynamics-fundamentals-new-paradigm/dp/1540583945

https://www.amazon.com/MoED-modification-electro-dynamics-priciple/dp/1537165690

CAN I ASK A FAVOUR? (USE A HEADING THAT WONT SHOW IN TOC)

If you enjoyed this book, found it useful or otherwise then I'd really appreciate it if you would post a short review on Amazon. I do read all the reviews personally so that I can continually write what people are wanting.

If you'd like to leave a review then please visit the link below:

(add the link to your kindle book. You will need to wait until the book is published before being able to get this link)

Thanks for your support!